CHEM 1120
Laboratory Manual

Preliminary Edition

Gary D. White
Middle Tennessee State University

KENDALL/HUNT PUBLISHING COMPANY
4050 Westmark Drive   Dubuque, Iowa 52002

Cover art courtesy of Liquid Library © 2005.

This edition has been printed directly from print-ready copy.

Copyright © 2005 by Gary D. White

ISBN 978-0-7575-2415-8

Kendall/Hunt Publishing Company has the exclusive rights to reproduce this work,
to prepare derivative works from this work, to publicly distribute this work,
to publicly perform this work and to publicly display this work.

All rights reserved. No part of this publication may be reproduced,
stored in a retrieval system, or transmitted, in any form or by any
means, electronic, mechanical, photocopying, recording, or otherwise,
without the prior written permission of Kendall/Hunt Publishing Company.

Printed in the United States of America

Dedication

To Mr. Ernest J. Scheidemann, my high school chemistry teacher

Contents

Preface……………………………………………………………………………………..vii

Acknowledgments……………………………………………………………………….ix

To the Student …………………………………………………………………………..xi

Laboratory Rules ………………………………………………………………………xiii

Safety Agreement ………………………………………………………………………xv

Experiment 11: Chromatography of M&Ms Candies………………………………………1

Experiment 12: Qualitative Analysis: Group I Cations……………………………………9

Experiment 13: Qualitative Analysis: Anions……………………………………………15

Experiment 14: Solubilities of Ionic Solids and Gases in Water………………………….21

Experiment 15: Freezing Point Depression………………………………………………35

Experiment 16: Concepts in Chemical Kinetics…………………………………………. 43

Experiment 17: Chemical Equilibrium……………………………………………………55

Experiment 18: Concentration of an Unknown Acid……………………………………..63

Pre-Lab Assignment for Experiment 19: Titration Curves………………………………..71

Experiment 19: Titration Curves………………………………………………………….75

Experiment 20: Solubility Product Constant of Magnesium Carbonate………………….83

Experiment 21: Electrochemical Cells……………………………………………………87

Literature Cited…………………………………………………………………………..95

Preface

The best method of learning chemistry is by active participation, whether in the classroom or laboratory. This laboratory manual enables students to study concepts important in the second semester of a General Chemistry course by doing experiments which focus on these concepts. Hopefully, through experimentation, concepts which may seem abstract or fuzzy in the textbook will become concrete and clear.

Acknowledgments

The author wishes to thank Professor Roy Clark for reviewing the first version of this laboratory manual. His thoughtful comments and suggestions are greatly appreciated.

Thanks also go to Professor Nabil Wakid who contributed some suggestions to improve the first version of this laboratory manual. I have incorporated these suggestions into the revised manual.

Parts of the kinetics and chemical equilibrium experiments began as lunchtime discussions between Professor John R. Luoma and the author. Sadly, Professor Luoma has passed away since the first version of this laboratory manual was written.

The author also wishes to thank Samantha Chesak for her work in testing the new solubility and kinetics experiments that appear in this manual.

Finally, the author wishes to thank his friends and family for their support and encouragement in helping me see this project to its completion.

To the Student

Welcome to Chemistry 1120 Laboratory! This laboratory manual contains 11 experiments. Each experiment contains a brief introduction which discusses the purpose of the experiment and and provides some useful background information. Please be sure to read these first few paragraphs before moving on to the procedure. There are sections in the text which are **boldfaced**. The boldfacing refers to tasks that you need to do on the report sheet. The boldfacing also indicates places which need special attention . The report sheet should be turned in to your lab instructor at the end of the lab period.

Some of these experiments contain elements of discovery learning. In discovery learning you proceed without knowing what to expect and then "discover" the concepts as you go along . It is a good way to learn because you see the concept develop as a result of doing the experiment. There are also some classic experiments which have been around for a long time like the qualitative analysis experiments. There is a wealth of information in the qual experiments. You can learn descriptive chemistry and develop analytical reasoning skills.

This laboratory manual is a revised version of the first edition of the laboratory manual originally titled Discovering Chemistry vol. 2. If you have comments and corrections please bring them to my attention. I hope that you enjoy the laboratory experiments.

Gary D. White,
Coordinator CHEM 1110 and 1120

Laboratory Rules[1]

All students are required to follow the laboratory rules listed. If additional safety rules are required, you will be notified either verbally or in writing. Violation of the rules may result in you being removed from the laboratory for the one lab period. Continual violation of the rules will be grounds for action aimed at removing the violator from the laboratory on a permanent basis.

1. Approved safety **goggles** will be worn at all times. Persons with contact lenses are encouraged to wear glasses to lab, if possible. Students wearing eyeglasses or contacts must wear safety **goggles** also. Approved safety goggles are available at Phillips Bookstore or from the Chemistry Club.
2. No smoking, chewing tobacco, or dipping snuff in the laboratory.
3. No eating or drinking in the laboratory. Any food or beverage brought into the laboratory will be considered contaminated and will be disposed of immediately.
4. No horseplay, roughhousing, messing around or any other form of physical activity not directly required for the experiments will be tolerated.
5. Students must have supervision while in the lab. You are not permitted to work in the laboratory without supervision.
6. No unauthorized experiments will be performed by any student.
7. No **open-toed shoes, sandals, or huaraches** (including Birkenstocks and Tevas) are to be worn in the laboratory.
8. Shorts, skirts and dresses must be **knee-length** or longer. All students are reminded that chemicals have a tendency to spill and to stain clothes, and students are encouraged to wear old clothes to lab.
9. No exposed midriffs (stomach area) will be permitted.
10. Long hair (shoulder length or longer) should be tied back so that it does not hang over open flames or in the student's face. This rule applies to both sexes equally.
11. All laboratory procedures, such as pipetting, will be followed. See the laboratory manual for appropriate procedures.
12. All accidents involving personal injury, no matter how minor the injury, must be reported to the Teaching Assistant, supervisor, or to the instructor **immediately.**
13. All accidents involving spilled chemicals or broken glass, no matter how minor, must be reported to the Teaching Assistant, supervisor, or to the instructor **immediately.**
14. The work area must be kept clean at all times.
15. Students must know the location of the eyewashes and safety showers.

Safety Agreement

Initial the blank preceding each numbered statement after you have carefully read it to indicate that you understand it and agree to abide by it. When you have initialled all items, complete the information at the bottom of the sheet.

_____ 1. I have received a copy of the laboratory rules, also listed on the back of this agreement. They have been explained to me by the laboratory instructor. I understand these rules and recognize that it is my responsibility to follow them at all times.

_____ 2. I recognize that my instructor may give me additional safety instructions, either verbally or in writing, during the course of the lab. I agree to follow these additional instructions and accept this as my responsibility,

_____ 3. I certify that, to the best of my knowledge, I do not have any of the following medical conditions except as indicated. I realize that the lab instructor will keep this information private and confidential unless it is necessary to reveal it to emergency personnel during a medical emergency.
    Note: Providing this information is optional
        ___allergies (list)
        ___asthma
        ___high blood pressure
        ___hypoglycemia
        ___diabetes
        ___pregnancy
        ___epilepsy
        ___hemophilia
        ___infectious blood diseases

_____ 4. I certify that (check one)
        ___I do wear contact lenses.
        ___I do not wear contact lenses.

_____ 5. I accept the authority of the lab instructor, the lab coordinator, or any official of the university (TA, staff, or faculty) to remove me from the laboratory if I violate this safety agreement.

Name (print)_____

University ID number_____

Date_____

Signature_____

Experiment 11
Chromatography of M& Ms candies

Goal of today's experiment: M&Ms candies are colored with organic dyes. The package states that the dye yellow #5 is one of the ingredients. Suppose you are allergic to this dye. Should you avoid eating all M&Ms candies or are there certain color M&Ms which are okay to eat? Which color M&Ms contain yellow #5? We'll use the technique of paper chromatography to answer this question.

Background: The word chromatography literally means the "writing of colors". It is a technique used to separate a mixture into its constituents. By comparing the constituents to known compounds they may be identified.

One kind of chromatography is **paper chromatography.** To help you understand how paper chromatography works consider the following analogy. Let's suppose there are three bicyclists traveling on the same road. One of these cyclists rides a mountain bike with fat tires. Another cyclist rides a road bike with skinny tires. And a third one rides a hybrid bicycle with tires whose width is larger than the road bike but smaller than the mountain bike. Across the width of the road at regular intervals there are iron gratings. The slats in the grating are parallel to the direction of the road . The spacing between the slats is about equal to the width of the hybrid bicycle tires. If the cyclist all start riding from the the same place on this road what will happen to the distance beween the cyclists as they proceed ?

The bike with skinny tires will get stuck in the grating making it more difficult for the cyclist to keep up with the other riders. The mountain bike with fat tires will ride easily over the grating. The rider will proceed smoothly over the road. The hybrid bicycle will may get stuck too although less so than the road bike. The distance beween riders will increase as they travel further on the road. Now let's replace this analogy with a real molecules. Instead of three cyclists we've got three different solutes in a solution. We place the solution on a piece of chromatography paper, the road. The paper and the solutes will experience different intermolecular forces between them. The solutes which are strongly attracted to the paper will tend to travel less quickly than the ones which are weakly attracted. So it is possible then to separate a mixture of solutes .

The R_f value. Figure 1 shows a developed chromatogram after the solvent has risen from bottom to top, carrying along with it the solutes. Some of the solutes lagged behind the others like bicyclists in a race. Each vertical column of spots originated at a sample spot directly below it on the lowest line.

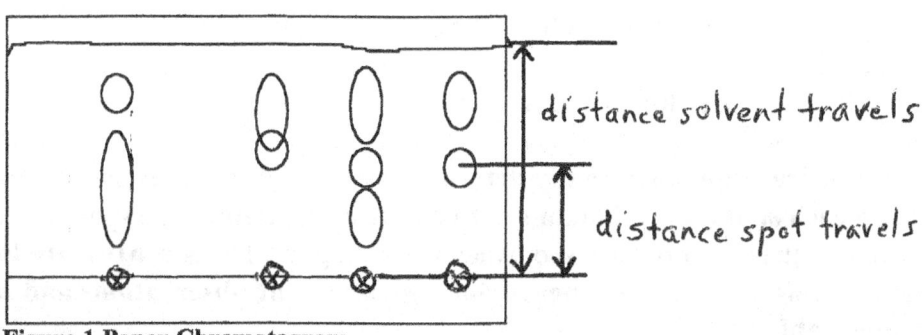

Figure 1 Paper Chromatogram

We may quantify the position of each constituent on the chromatogram by computing the retention factor or R_f value. We define this as the distance the solute travels divided by the distance the solvent travels. The R_f values may be used to characterize and hence identify the components in a sample.

$$R_f = \frac{\text{distance spot travels}}{\text{distance solvent travels}}$$

If 2 spots have the same R_f value then they may be the same substance.

Procedure: You may work with a lab partner. Obtain a ceramic spot plate from the cart. Place 1 M&M candy of each different color into a well in the spot plate. Place 1 drop of DI water on top of each candy. Using a different toothpick for each color flip each of the candies over in the well. The water in the well should take on the color of the candy.

Continue flipping the candy in the well to obtain as deep a color as possible. The chromatagraphy paper supplied has dimensions of about 8 cm X 10 cm. The fibers in the paper are aligned parallel to the shorter dimension. It is important to use the paper so that the shorter dimension is vertical as shown in the Figure 1. Draw a straight line 10 cm long **in pencil** about 1 cm from the edge.

Lining up for the race: Make 7 small "X"s in pencil on the line spaced about 1 cm apart from each other beginning about 1 cm from the edge. These Xs are the starting blocks, one for each sample. Dip the tip of a toothpick into one of the wells. Lightly touch the tip to one of the pencil marks on the chromatography paper. Try to make the spot no larger than the this o. Let the spot dry. Repeat this step 2 more times to increase the amount of dye in the spot. Repeat the procedure for the other colored M&Ms but **skip the pencil mark in the middle of the paper.**

Put 2 drops of yellow dye #5 into an empty well in the spot plate. Dip the tip of a toothpick into the dye and make a spot on the pencil mark you previously skipped. This is the solvent only lane.

Let the all the spots dry **completely**. You may accelerate the drying process by blowing on the spot or using compressed air.

Developing the chromatogram: Roll the paper into the shape of a cylinder and staple the bottom and the top as shown leaving a little bit of space between the ends. The filter paper should now be able to stand up by itself in a 400 ml beaker without touching the walls of the beaker. Remove the filter paper from the beaker. Pour some 0.1% NaCl into the 400 ml beaker to a depth of 4-5 mm.

Starting the race: Carefully place the filter paper in the beaker. The colored spots should be close to **but not touching** the solution. Through capillary action the solvent will rise through the filter paper. **DO NOT DISTURB** the developing chromatogram.

Stopping the race: When the solvent reaches about 2 cm from the top of the filter paper carefully remove the filter paper from the beaker. Remove the tape and lay the chromatogram flat on the desk.

Judging the race: For each candy determine which colors are present. **Measure the distance each spot travels in mm relative to the pencil line.** The spots will be oblong and may overlap. Use the center of each spot when measuring the distance. **Measure the distance the solvent travels relative to the pencil line. Record your observations and measurements in the Data Table.**

Calculations

Calculate the R_f value for each spot using the equation given on the previous page. Record the results in the Results Table.

Name	Date
Lab Partner	Lab Instructor

Experiment 11
Chromatography of M&Ms Candies

DATA TABLE

Sample	Color of Spot / Distance Spot Travels in mm					Distance Solvent Travels mm
Yellow candy	-------	-------	-------	-------	-------	
Orange candy	-------	-------	-------	-------	-------	
Green candy	-------	-------	-------	-------	-------	
Yellow #5	-------	-------	-------	-------	-------	
Blue candy	-------	-------	-------	-------	-------	
Brown candy	-------	-------	-------	-------	-------	
Red Candy	-------	-------	-------	-------	-------	

DATA TABLE

Sample	Color of Spot / R_f factor				
Yellow candy	-------	-------	-------	-------	-------
Orange candy	-------	-------	-------	-------	-------
Green candy	-------	-------	-------	-------	-------
Yellow #5	-------	-------	-------	-------	-------
Blue candy	-------	-------	-------	-------	-------
Brown candy	-------	-------	-------	-------	-------
Red Candy	-------	-------	-------	-------	-------

Show one sample calculation of the R_f factor

Food for thought.

1. Based upon your results is Yellow Dye #5 found in M&Ms candies . If so which colors? Explain your answer.

2. Based upon your results are there any M&Ms candies which share any other color dyes?

Attach your chromatogram here. Write your name(s) in pencil on the chromatogram.

Experiment 12
Qualitative Analysis : Group I Cations

Introduction: This experiment is the first of two experiments which focus on the subject of **qualitative analysis**. In qualitative analysis the identity of the constituents in a sample is determined. For these experiments the constituents are ions. In order to identify the ions we will separate them. We can do this because the chemical reactions of common ions with different reagents are well known. For example it is well known that Ag^+ ions react with Cl^- ions to form an AgCl precipitate. AgCl is white and is insoluble in hot water. So if we find a white precipitate which doesn't dissolve in hot water it's possible that it could be AgCl.

As another example, suppose Ag^+ ions are found in a solution which also contains Na^+ ions. By adding Cl^- ions the Ag^+ ions can be removed as AgCl leaving the Na^+ behind in solution. (Remember NaCl is very soluble) Thus we have a method by which we can separate and identify Na^+ ions from Ag^+ ions.

In this particular experiment we'll be identifying one or more of the following ions Ag^+, Pb^{2+}, Hg_2^{2+}. The species Hg_2^{2+} may appear strange to you. Hg^+ is unusual in that it dimerizes to form Hg_2^{2+} and thus appears to be a 2+ ion. The unknown solution may contain one of these ions, a mixture of any two of these ions or perhaps all three of these ions.

Procedure: The procedure is divided up into two parts. First you'll work with some known solutions to understand the important underlying chemistry and what to expect. Then you'll do experiments with an unknown sample. **For Part I you may work with a lab partner. Part II requires that you work individually.**

Part I. Preliminary experiments: Put three small test tubes in a test tube rack. Place 1 mL of each of the following solutions into the test tubes.
$AgNO_3(aq)$
$Pb(NO_3)_2(aq)$
$Hg_2(NO_3)_2(aq)$
Label the test tubes with either a pencil or by keeping track of their positions in your test tube rack.

Add 10 drops 6 M HCl(aq) to each test tube. **Describe the changes which occur, if any. For each chemical change write the corresponding balanced molecular equation.**
Set up a water bath using a 400 mL beaker. Heat the water in the bath to boiling and continue heating. Put the three test tubes in the boiling water. Use a stirring rod to break up any precipitates which formed into smaller pieces. Heat the test tubes in boiling water for 5 minutes, stirring the solution occasionally. **Describe the chemical changes which occurred if any.**

Carefully add 10 drops K_2CrO_4(aq) to the test tube containing Pb^{2+} ions while the test tube is still hot. Now remove the test tubes from the bath. **Describe the chemical change. Write the corresponding molecular equation.** To the remaining two test tubes add enough 6 M NH_3(aq) until the solution is basic. Use red litmus paper to determine if the solution is basic. Be sure the contents of the test tube has been thoroughly mixed. Then dip the stirring rod into the mixture. Now touch the stirring rod to the litmus paper. Do not dip the litmus paper directly into the mixture. Add 10 more drops 6 M NH_3(aq) to the remaining two test tubes. Stir the contents of each test tube vigorously. **Describe the chemical changes.** To the solution containing the Ag^+ ions add 6M HCl(aq) until the solution is acidic. **Describe the chemical changes.**

Part II. Unknown

The flow chart on the following page refers to the steps given below.

Step 1. Obtain a Group I* unknown sample from your instructor. **Record the unknown number.** Put about 3 mL of the Group I unknown solution into a small test tube. Add 20 drops of 6 M HCl(aq) and stir the solution with a stirring rod. Centrifuge for about 1 minute. Remove the test tube from the centrifuge. Add several more drops of 6M HCl(aq). Check for complete precipitation by observing the reaction of HCl(aq) with the supernatant. If you see more precipitate form continue adding HCl(aq). When the addition of HCl(aq) no longer forms a precipitate stop adding HCl. Decant the liquid.

Step 2. Add 3 mL H_2O to the precipitate and place the test tube in a boiling water bath. While the test tube is heating use your glass stirring rod to break up the solid into smaller pieces. Heat for about 5 minutes, stirring occasionally. While the test tube is hot, centrifuge for about 45 seconds. Immediately decant the liquid into a small test tube.
Do not discard the liquid.

Step 3. If your sample contains $PbCl_2$ it will have dissolved in the hot water. Add 10 drops of K_2CrO_4(aq) to the **liquid**. A **yellow** precipitate confirms the presence of Pb^{2+} ions. If you are uncertain about whether you have a precipitate, centrifuge to see if a solid appears on the bottom of the test tube.

Step 4. If a precipitate remains after you decanted your sample in Step 2 you may have AgCl or Hg_2Cl_2 or both. Add 3 mL NH_3(aq) to the precipitate. Vigorously stir the sample with a stirring rod. Check to see if the solution is basic with red litmus paper. If it is not add more NH_3(aq) until the solution is basic. Centrifuge and decant the liquid into another test tube. A black residue (solid) indicates the presence of Hg^{2+}. To the liquid add 6 M HCl(aq) until the solution is acidic. A white precipitate confirms the presence of Ag^+ ion.

*The designation Group I does not mean refer to Group IA or IB on the periodic table. The name means it's the first group of ions to forms precipitates when we separate ions according to the kind of precipitates that are formed when various reagents are added.

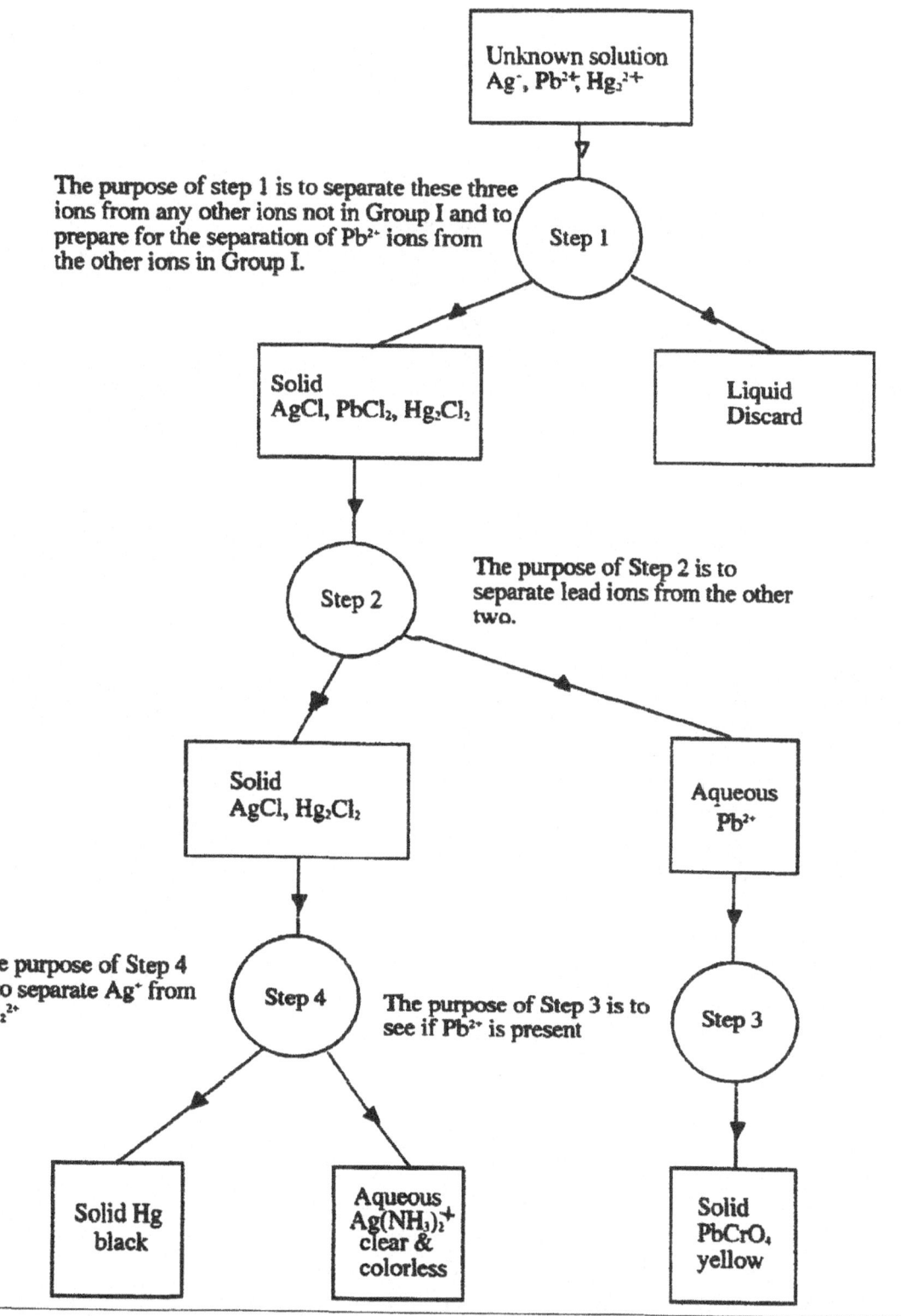

Name	Date
Lab Partner for Part I	Lab Instructor

Experiment 12
Qualitative Analysis : Group I Cations

Part I. Preliminary experiments

	Description of chemical changes which occur upon addition of 6 M HCl
$AgNO_3$(aq)	
$Pb(NO_3)_2$	
$Hg_2(NO_3)_2$	

	Balanced molecular equation
$AgNO_3$(aq)	
$Pb(NO_3)_2$	
$Hg_2(NO_3)_2$	

Description of chemical changes which occur upon heating in boiling water bath

Describe chemical change and write corresponding molecular equation for when K_2CrO_4(aq) is added to test tube containing Pb^{2+} ions

Describe chemical changes which occur when 6 M NH_3 (aq) is added.

Describe chemical change which occurs when 6 M HCl(aq) is added

Part II. Unknown

Decide which ions are present in your unknown sample. Fill out the form below and return it to your instructor. If you are correct you will receive 5 points. If you are incorrect you get a second chance. You receive 4 points if you are correct on the second try. You receive -1 point for each error.

Qualitative Analysis: Group I Unknown

Name
Unknown Sample #

Ions Present _____

Score

First Report	0 errors	5 pts
Second Report	0 errors	4 pts
	1 error	3 pts
	2 errors	2 pts
	3 errors	1 pt

Lab Instructor's Initials _____

Experiment 13
Qualitative Analysis : Anions

Introduction: In this experiment we'll identify anions contained in an unknown aqueous solution. The approach to identifying anions is different than the one we used for cations. For the cations we used a flowchart separation scheme, carrying our results from one step to the next. In contrast, we will do **spot tests** for anions. Spot tests are separate tests in which we'll use a fresh sample of unknown solution for each test.

The spot tests described below are written for a **known** solution. The known solution is **boldfaced**. You should first do the test on the known solution, answer any questions asked and then repeat the test on your unknown solution.

Procedure: Work with a lab partner on the known solutions. Work individually on the unknown solution.

Test #1: Carbonate

Obtain a rubber stopper and bent delivery tube from the cart. Put about 3 mL of **Na_2CO_3(aq)** into a test tube which will fit the rubber stopper. In a medium size test tube put about 10 mL of limewater, $Ca(OH)_2$(aq). The limewater must be a clear and colorless solution. If it is cloudy centrifuge the limewater and decant the liquid. Place the rubber stopper and delivery tube onto the test tube containing the Na_2CO_3(aq) solution. Put the end of the delivery tube into the limewater.

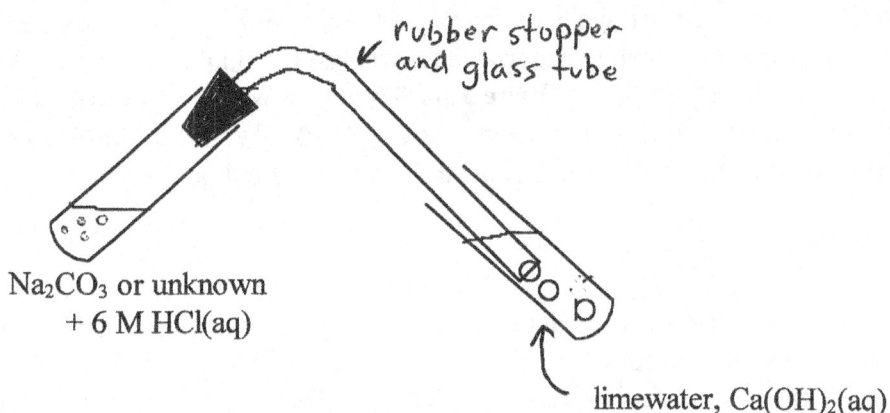

Remove the stopper and add a few drops of 6 M HCl(aq) to the **Na_2CO_3(aq)** solution. **Immediately** replace the stopper and check for the evolution of bubbles. The limewater will turn cloudy if CO_3^{2-} is present.
Identify the bubbles. Write a balanced equation which shows the reaction of Na_2CO_3(aq) with HCl(aq) Identify the precipitate. Write a balanced equation or equations to show a plausible way to form the precipitate. Hint: The precipitate is **chalky** white.
Repeat the above procedure with your unknown.

Test #2: Nitrate

CAUTION: Exercise extreme caution when performing this test. Concentrated sulfuric acid causes severe burns. Make sure you are wearing your goggles!

Prepare 5 mL of a saturated solution of $FeSO_4(aq)$ by putting a small amount of solid $FeSO_4$ about the size of a pea into a small test tube half-filled with DI H_2O. Stir thoroughly. If a small amount of solid remains on the bottom of the test tube then the solution is saturated. Add more $FeSO_4$ as necessary until the solution is saturated. In another small test tube put about 20 drops of **$NaNO_3$(aq). In the hood carefully and slowly** add 20 drops of concentrated H_2SO_4 to the **$NaNO_3$ with the test tube pointing away from yourself and others.** Stir the solution thoroughly. Cool the test tube in some ice water. Remove the test tube from the ice water and tilt it at about a 45 degree angle with the mouth pointing upward. Slowly pour about 2 mL of the saturated $FeSO_4$(aq) solution down the side of the test tube as shown in the figure.

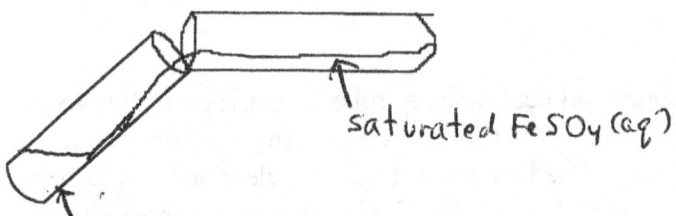

$NaNO_3(aq)$ or unknown + 20 drops concentrated $H_2SO_4(aq)$

A brown ring formed at the interface between the solutions confirms the presence of NO_3^-. Hold the test tube against a piece of white paper to help you see the ring more clearly. In this test hydronium ion oxidizes Fe^{2+} to Fe^{3+}. **Write a half-reaction which shows the loss of electrons from Fe^{2+} to form Fe^{3+}.** The NO_3^- ion is reduced to NO(g). **Write the half-reaction which shows this reduction.** Repeat the above procedure with your unknown.

Test #3: Chloride

Put about 20 drops **NaCl(aq)** in a small test tube. Acidify the solution with 6 M HNO_3, if it is not already acidic. Add several drops of $AgNO_3$(aq). A white precipitate indicates the presence of Cl^-. **Write a balanced chemical equation which describes the reaction between NaCl(aq) + $AgNO_3$(aq).** Why do you think it might be necessary that the solution be acidic before adding $AgNO_3$(aq) ? Hint: What if $AgCO_3$ precipitate had been present?

Test #4: Phosphate

Put 20 drops of **Na_3PO_4(aq)** solution in a small test tube. Add 20 drops 6 M HNO_3 and 40 drops of ammonium molybdate solution, $(NH_4)_2MoO_4$(aq). Heat in a warm water bath for a few minutes. Remove the test tube and let stand for at least 20 minutes. A yellow precipitate confirms the presence of PO_4^{3-}. **What element is responsible for the yellow color of the** precipitate ? Hint: What color are compounds formed from main group elements ? What color is the precipitate? Repeat the above procedure with your unknown .

Test #5: Sulfate

Put 20 drops **Na_2SO_4(aq)** solution in a small test tube. Add 20 drops 6 M HCl(aq) Add 20 drops $BaCl_2$(aq) solution. A white precipitate indicates that SO_4^{2-} is present. **Write a balanced chemical reaction which describes the formation of the white precipitate from Na_2SO_4(aq) and $BaCl_2$(aq).** Repeat the above procedure with your unknown .

Complete the report sheet and turn it in to your instructor.

Name	Date
Lab Partner for known solutions	Lab Instructor

Experiment 13
Qualitative Analysis: Anions

Test #1: Carbonate

The bubbles are

Write a balanced equation which shows the reaction of $Na_2CO_3(aq)$ with $HCl(aq)$

The precipitate formed in the limewater is

Write a balanced equation or equations which show a plausible way the precipitate is formed

Test #2: Nitrate

Write a balanced half-reaction which shows the loss of electrons from Fe^{2+} to form Fe^{3+}

Write a balanced half-reaction which shows the reduction of $NO_3^-(aq)$ to form $NO(g)$

Test #3: Chloride

Write a balanced chemical equation that describes the reaction between $NaCl(aq) + AgNO_3(aq)$

It might be necessary to acidify the solution before adding $AgNO_3$ because

Test #4: Phosphate

The element most likely responsible for the yellow color of the precipitate is

Test #5: Sulfate

Write a balanced molecular equation which describes the formation of the white precipitate from Na_2SO_4(aq) and $BaCl_2$(aq)

Decide which ions are present in your unknown sample. Fill out the form below and return it to your instructor. If you are correct you will receive 5 points. If you are incorrect you get a second chance. You receive 4 points if you are correct on the second try. You receive -1 point for each error.

Qualitative Analysis: Anions

Name
Unknown Sample #

Ions Present _____

Score

First Report	0 errors	5 pts
Second Report	0 errors	4 pts
	1 error	3 pts
	2 errors	2 pts
	3 errors	1 pt

Lab Instructor's Initials _____

Experiment 14
Solubilities of Ionic Solids and Gases in Water

Introduction: The **solubility** of a substance is the maximum amount of solute that dissolves in a given quantity of solvent at a given temperature. In this experiment, you'll compare the solubilities of two ionic solids NaCl and $PbCl_2$ in water. We'll see how the solubilities of these substances change with temperature. Then we'll examine how the solubility of carbon dioxide gas in water depends on temperature.

Predictions: In the space provided on your report sheet briefly state what you believe to be the effect that temperature has on the solubility of a solid in water. Then state what you believe to be the effect that temperature has on the solubility of a gas in water.

Procedure: Work with a lab partner. You and your partner will collect data about the solubility of NaCl and $PbCl_2$ at a given temperature. Your lab instructor will assign you two temperatures, one for each of the solids.. Then we'll pool our data together to study the effect that temperature has on the solubility.

The solubility of NaCl in H_2O: First prepare a clean and dry evaporating dish. Clean the evaporating dish using a little soap, water and gentle scrubbing with a test tube brush. Dry the dish using a Bunsen burner flame. Let the dish cool to room temperature on a wire gauze pad. When the dish is cool to the touch weigh it to the nearest .01 g using the balance. **Record the mass.**

Set up a water bath by filling an 800 mL beaker about three-fourths full of tap water. Place the beaker on a wire gauze pad supported by a ring or tripod. Heat up the water to the assigned temperature. Do your best to keep the temperature in the water bath at this temperature. You'll need to measure the temperature periodically. If the water gets too hot, discontinue heating. If it starts to cool, begin heating again. If you are assigned room temperature then you will not need to heat up the water but since the dissolving of a solid may be exothermic, use the water bath to keep the temperature constant. Measure 5.0 mL of deionized water into a medium size test tube. Place the test tube in the water bath. Measure the temperature of the water in the bath and the water in the test tube. When the two temperatures are about the same, weigh out 2 grams of NaCl on a piece of weighing paper to the nearest .01 g and add it to the test tube. Carefully stir the contents of the tube in the water bath with a stirring rod for at least 10 minutes and observe what happens. **Does all the solid dissolve?** While you or your lab partner are stirring the test tube the other person should set up a funnel on a ring stand or tripod. Fold a piece of filter paper in half and then in half again. Tear off a small corner of the paper and open the paper up so that it fits inside the funnel. Place a small beaker (10 or 50 mL) underneath the funnel. Now quickly remove the test tube from the water bath and filter the contents through the filter paper. The liquid in the beaker should be clear. Transfer the liquid to the evaporating dish. **Weigh the liquid and dish on the balance.** Record the mass. Now heat the liquid **gently** until the all the water has evaporated and crystals of NaCl are left behind on the dish. Do not heat the solid too rapidly as spattering may result which may burn your skin and also result in loss of product. When the dish is cool to the touch weigh the dish and recovered NaCl on the balance. Continue heating and weighing until a constant weight to within .02 g is obtained. Record the mass of the dish plus product Calculate the mass of the solution and the mass of the solid NaCl. Now calculate the mass of the water in your solution. Calculate the mass of NaCl per 100 grams of water. Report this value and the temperature to your instructor. To dispose of the NaCl add enough tap water to your evaporating dish until the solid dissolves entirely. Empty the contents of the dish in the sink. Rinse the solid NaCl residue on the filter paper off with tap water into small beaker. Use enough water until all the solid dissolves. Empty the beaker into the sink.

The solubility of $PbCl_2$ in H_2O: First we'll make the $PbCl_2$ solid by reacting an aqueous solution of lead nitrate, $Pb(NO_3)_2$(aq) with HCl(aq). Write the balanced molecular equation which describes the reaction between $Pb(NO_3)_2$(aq) and HCl(aq). Add 2 mL of $Pb(NO_3)_2$(aq) and 4 mL of HCl(aq) to a small test tube. Stir the contents with a stirring rod. A white precipitate will form in the test tube. Place the test tube in the centrifuge together with a matching blank test tube and centrifuge for about 1 minute. Pour off the liquid into a waste container in the hood. Add a few mL of cold deionized water to the test tube. Stir the contents with your stirring rod. Centrifuge again for about 1 minute. Pour off the liquid waste into the appropriate waste container. Using 10 mL of deionized water transfer the precipitate in the test tube to a medium size test tube. Using the water bath you set up in the previous part, heat the water to the assigned temperature. Do your best to keep the temperature constant. Place the medium size test tube in a water bath and allow the contents of the test tube and the water bath to equilibrate. Carefully stir the contents of the tube in the water bath for at least 10 minutes and observe what happens.

Does all the solid dissolve? While you or your lab partner are stirring the test tube the other person should set up a funnel on a ring stand or tripod. Fold a piece of filter paper in half and then in half again. Tear off a small corner of the paper and open the paper up so that it fits inside the funnel. Place a small beaker (10 or 50 mL) underneath the funnel. Now quickly remove the test tube from the water bath and filter the contents through the filter paper. The liquid in the beaker should be clear. Transfer the liquid to the evaporating dish. **Weigh the liquid and dish on the balance. Record the mass.** Now heat the liquid **gently** until the all the water has evaporated and crystals of $PbCl_2$ are left behind on the dish. Do not heat the solid too rapidly as spattering may result which may burn your skin and also result in loss of product. When the dish is cool to the touch weigh the dish and recovered $PbCl_2$ on the balance. Continue heating and weighing until a constant weight to within .02 g is obtained. **Record the mass of the dish plus product. Calculate the mass of the solution and the mass of the solid $PbCl_2$. Now calculate the mass of the water in your solution. Calculate the mass of $PbCl_2$ per 100 grams of water. Report this value and the temperature to your instructor.** To dispose of the $PbCl_2$ product add a little tap water to the dish and transfer the contents of the dish to the waste container in the hood. Rinse the filter paper with some tap water into a small beaker and transfer the contents of the beaker to the waste container. Do not put the filter paper into the waste container.

Solubility of carbon dioxide in H_2O: In this part you'll determine the effect of temperature on the solubility of a gas in a liquid. Seltzer water contains carbonated water. Carbonated water is made by dissolving carbon dioxide gas in water. **Write the**

chemical reaction, which describes this process. Hint: the acid formed is carbonic acid.

We'll determine how much carbon dioxide dissolves by reacting the carbonic acid with just enough aqueous sodium hydroxide, NaOH until the reaction is complete. **Write the chemical reaction which describes the reaction between NaOH(aq) and carbonic acid. Carbonic acid is unstable. Now rewrite this equation but break up the carbonic acid into two small familiar molecules.** Use a graduated cylinder to measure out 25.0 mL of room temperature seltzer water into a 150 ml beaker. Next you'll need to calibrate a glass dropper. Using some tap water in a small container, count out 40 drops into your 10 mL graduated cylinder. Be sure to hold the dropper vertically. **Measure the volume occupied by 40 drops to the nearest 0.1 mL. Record the result. Calculate the volume per drop.** Add 2 drops of phenolphthalein indicator to the seltzer water in your beaker. Place a few mL of 1.0 M NaOH(aq) into a small beaker. Add drop wise 1.0 M NaOH to the seltzer water. Gently swirl the flask after each addition to mix the contents. Continue adding NaOH dropwise until the solution turns pink and the pink persists for 30 seconds or more. **Record the number of drops necessary to achieve this pink color. Record the temperature of the solution. Calculate the volume of NaOH delivered in mL. Convert the number of mL of NaOH to number of moles of NaOH. Using the reaction stoichiometry determine how many moles of carbon dioxide is required to react with the number of moles of NaOH you added. Convert the number of moles of CO_2 to grams. Now assuming that the density of carbonic acid is about the same as the density of water, calculate the mass of your sample in grams. What is the weight of water in your sample? Report the solubility of CO_2 in g/100 g H_2O.**

Using the water bath from the previous parts of this experiment raise the temperature of your sample to about 20 °C above room temperature. Measure 25.0 mL of seltzer water into a large test tube and heat in the water bath until the temperature inside the test tube reaches the temperature of the bath. Heat an additional 5 minutes, maintaining the temperature. Remove the test tube from the bath and empty the contents into a 150 ml beaker. Add NaOH dropwise as you did before until the solution turns pink for at least 30 seconds. **Record the temperature. In the same manner as you did before determine the solubility of CO2 in g/100g.**

Pooling the data and plotting the results:

Your lab instructor will post the results of the solubilities of NaCl and $PbCl_2$ on the blackboard. **Copy the pooled data into your report sheet. On the graph paper plot the solubility of NaCl in g/100g H_2O versus temperature in °C. If the temperatures are the same for more than one group of students you should compute the average solubility. If the temperatures are different then plot the points separately.** Choose an appropriate scale for the x and y-axis so that the graph is spread out over the graph paper. **Make a small circle around each data point. Connect the points you have drawn with a line. Repeat the graphing procedure for $PbCl_2$ on the second piece of graph paper.**

Name	Date
Lab Partner	Lab Instructor

Experiment 14
Solubilities of Ionic Solids and Gases in Water

Predictions: Briefly state what you believe to be the effect that temperature has on the solubility of a solid in water. Then state what you believe to be the effect that temperature has on the solubility of a gas in a liquid.

Solubility of NaCl and $PbCl_2$ in H_2O

	NaCl	$PbCl_2$
Assigned Temperature in °C		
After stirring for 10 minutes does all the solid dissolve?		
Weight of empty evaporating dish		
Weight of evaporating dish and solution		
Weight of evaporating dish and recovered solid		
Mass of solution Show calculation.		
Mass of solid Show calculation.		
Mass of H_2O Show calculation		
Mass of solid/100 g H_2O Show calculation		
Measured Temperature in °C		

Solubility of CO_2 gas in H_2O

Write the balanced molecular equation which describes the reaction between $Pb(NO_3)_2(aq)$ and $HCl(aq)$

Write the balanced molecular equation which describes carbon dioxide which describes carbon dioxide gas dissolving in water to form carbonic acid.

Write the chemical reaction which describes the reaction between $NaOH(aq)$ and carbonic acid.

Rewrite the equation which describes the reaction between $NaOH(aq)$ and carbonic acid but break up the carbonic acid into two small familiar molecules.

Volume in mL of 40 drops of H_2O

Volume per drop Show calculation

	Room temperature (RT)	Above RT
Temperature of solution in °C		
Number of drops necessary to achieve pink color		
Volume of NaOH delivered in mL		
Moles of NaOH		
Grams of CO_2		
Mass of Seltzer Water in grams. Assume density = 1.0 g/mL		
Mass of water in sample		

	RT	Above RT
Solubility of CO_2 in g/100 g H_2O		

NaCl Pooled Data

Team	Temperature in °C	Solubility in g/100 g H_2O

PbCl₂ Pooled Data

Team	Temperature in °C	Solubility in g/100 g H₂O
Team	Temperature in °C	Solubility in g/100 g H₂O

Questions:

1. Based upon your graphs what is the trend in the solubility of each of these compounds as the temperature of the solution increases. Is the temperature dependence the same for both compounds.

2. For a given temperature how does the solubility of NaCl compare to the solubility of $PbCl_2$?

3. Discuss why you think the solubilities of the two compounds are different. Try to think on a molecular level about the intermolecular forces involved.

4. Based upon your results in the second part of this experiment what do you conclude about how the solubility of a gas in a liquid changes with temperature?

5. What is Le Châtelier's principle?

6. If the dissolving of a gas is an exothermic process, explain in terms of LeChâtelier's Principle the relationship between the solubility of a gas in a liquid and temperature.

Experiment 15
Freezing Point Depression

Note: You will need a watch with a second hand or digital display for this experiment.

Purpose: This experiment illustrates one of the **colligative properties** of solutions, namely **freezing point depression**. You'll use this property to determine the molar mass of an unknown compound.

Introduction: You've probably had experience with freezing point depression in your everyday life. We add antifreeze to an automobile radiator to prevent the water in the radiator from freezing and expanding when the outside temperature dips below 0 °C. Antifreeze works by lowering the freezing point of water. The lowering of the freezing point of the solution relative to the freezing point of the pure solvent is known as **freezing point depression**. Quantitatively, we may express the freezing point depression as the difference between the freezing point of the pure solvent minus the freezing point of the solution or $\Delta T = T_f^0 - T_f$ where ΔT is the freezing point depression, T_f^0 is the freezing point of the pure solvent and T_f is the freezing point of the solution. Since T_f^0 is larger than T_f, ΔT is a positive quantity.

Freezing point depression depends on the number of solute particles present in the solution but not the nature of the solute particles. Thus freezing point depression depends on the concentration of the solution. A useful way to express concentration is in terms of molality represented by m.

$$m = \frac{\text{mol solute}}{\text{kg solvent}} \quad (1)$$

We find that the freezing point depression is directly proportional to the molality of the solution. For non-electrolytes this relationship can be expressed by

$$\Delta T = K_f m \quad (2)$$

where m is the concentration of the solution expressed in units of molality. K_f is the freezing point depression constant which depends on the solvent. Water has a K_f value equal to 1.86 °C/m In this experiment para-dichlorobenzene is chosen as the solvent because it has a freezing point which is above room temperature. Data tables indicate that the K_f for para-dichlorobenzene is 7.1 °C /m.

Procedure: Work with a lab partner. Obtain an unknown sample from your instructor. **Record the unknown number.** On a piece of paper weigh out about 2.0 grams of the unknown to the nearest .01 g. **Record the mass.** Set aside the unknown sample for now. Set up a water bath using a 600 ml beaker about 3/4ths full of water. Begin heating the water with a Bunsen burner.

Weigh a large test tube which is clean and dry to the nearest .01 g. **Record the mass.** Weigh out about 20 g of para-dichlorobenzene (PDB) to the nearest .01 g on a piece of paper. Carefully transfer the PDB into the large test tube.

Using a test tube holder or a clamp place the test tube containing the PDB in your water bath and heat the sample. When the sample begins to melt place a thermometer in the test tube.

Continue heating the sample until it completely melts. When the temperature reaches about 60 °C discontinue the heating process by removing the test tube from the water bath. **DO NOT HEAT THE SAMPLE ABOVE 60 °C.**

The sample will begin to cool. **Record the temperature every 30 seconds.** Gently stir the sample with the thermometer while the sample cools. Watch for the point at which the first bit of liquid freezes. **When this happens take 10 more temperature readings.** Then place the sample back into the water bath and warm the sample until it melts completely and the temperature reaches about 60 °C.

Add the weighed unknown sample to the PDB in the large test tube. Carefully stir the sample with the thermometer. When the sample is completely melted discontinue the heating process by removing the test tube from the water bath. The sample will begin to cool. **Record the temperature every 30 seconds.** Watch for the point at which the first bit of liquid freezes. **When this happens take 10 more temperature readings.**

Calculations: On the same graph, plot temperature in °C versus time in seconds for both samples. Your graph should look something like figure 1. The temperature axis should not start with 0 °C but be 1 or 2 °C lower than the lowest recorded temperature reading. **Using a ruler, draw the best straight line through the data points as the sample first begins to cool. Draw the best straight line through the data points beginning from the last of the data points.** Draw a horizontal line from the intersection to the temperature axis. This is the freezing point of the liquid. In the case of the solution you may notice a few data points which appear to dip below the freezing point. This phenomenon is called **supercooling** and should be ignored when determining the best line. **Report the freezing points for the pure solvent and the solution.**

Calculate the freezing point depression ΔT. Combine equation (1) and (2) to obtain an algebraic expression for the mol solute. Now use this expression to calculate mol solute. **Since mol solute = gram solute/ molar mass solute you should now be able to solve for the molar mass of the unknown. Calculate the molar mass of the unknown in units of g/mol.**

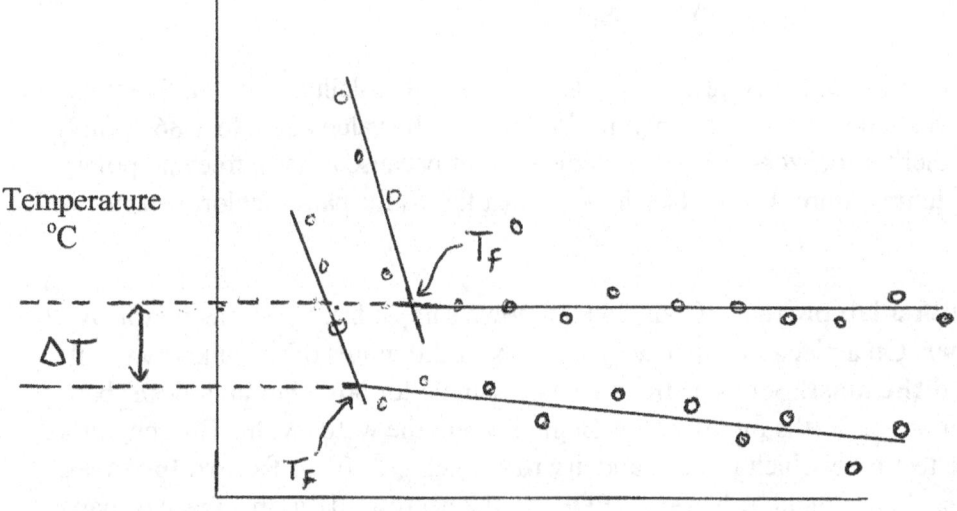

Figure 1 Temperature vs. time

Name	Date
Lab Partner	Lab Instructor

Experiment 15
Freezing Point Depression

Mass of PDB	Unknown Sample #	Mass of unknown

Data - PDB

Temperature °C	time (s)	Temperature °C	time (s)

Freezing point for PDB, T_f^o

Data- PDB + unknown

Temperature °C	time (s)	Temperature °C	time (s)

Freezing point for solution, T_f

$\Delta T_f =$

Derive expression for mol solute

Calculate mol solute

Calculate molar mass of unknown, M_m

Depressing Questions

1. How does the freezing point of a 0.100 m $BaCl_2$(aq) solution compare to the freezing point of a 0.100 m HNO_3(aq) solution? Explain your answer.

2. Was the unknown compound most likely ionic or organic? Explain.
   Hint: What kind of compound is the solvent, para-dichlorobenzene, $C_6H_4Cl_2$?

3. How many grams of the non-electrolyte ethylene glycol, $C_2H_6O_2$ must we add to 1264 grams of H_2O to lower the freezing point to -5.00 °C ? The freezing point depression constant for water is 1.86 °C/m .

Experiment 16
Concepts in Chemical Kinetics

Introduction: Chemical kinetics is the study of the speeds or rates of chemical reactions. You see examples of kinetics often in your everyday life. For example, a 25 lb. turkey will cook faster at a temperature of 350 °F than at 325 °F. As another example, when a dentist prepares an amalgam he will use a catalyst which will cause the amalgam to set quickly.

There are several factors which may affect the rate of a chemical reaction. In Part I of this experiment you'll discover some of these effects. Part II will introduce you to the concept of the half-life of a chemical reaction.

Part I. What factors affect the rate of a chemical reaction?

In this part we'll measure the speed of a chemical reaction and how the concentration of a reactant affects the speed of the reaction. The reaction between magnesium metal and hydrochloric acid is

$$Mg(s) + 2 HCl(aq) \rightarrow MgCl_2(aq) + H_2(g)$$

The net ionic equation is $Mg(s) + 2 H^+(aq) \rightarrow Mg^{2+}(aq) + H_2(g)$

One way to express the speed is by writing the rate law. The rate law is written as the product of concentrations of reactants each raised to an exponent called the order. For this reaction the rate law would be rate=$k[H^+]^m$. You may be wondering what happened to the concentration of magnesium metal in the rate law. How come it isn't in the rate law? It's a reactant isn't it? Since magnesium is a solid, its concentration is approximately constant. Thus its concentration doesn't appear directly in the rate law. The orders in a rate law are often integers, usually equal to one or two. We are going to determine whether the rate depends on the first or second power of the H^+ concentration.

Procedure: **Work with a lab partner.** Measure 25.0 mL of 4.0 M HCl(aq) in a 25.0 mL graduated cylinder. Divide the volume of acid in half by pouring 12.5 mL into a medium-sized test tube. **Save the remaining acid.** Set up a water bath by filling a large beaker (600 or 800 mL) about two-thirds full of tap water. Place the beaker on a wire gauze pad which rests on a ring attached to a ring stand. Place a thermometer in the water bath. Measure the temperature in °C. Record the result. Place the test tube in the water bath. Remove the thermometer from the water bath, wipe it dry with a paper towel and place it in the test tube. Record the temperature. While thermal equilibrium is being established between the acid and the water bath, measure and cut seven 1.0 cm strips of magnesium from the magnesium stock provided. Thermal equilibrium will be reached when the temperature of the liquid in the test tube is the same as the water bath temperature. While your partner times the event add one of the magnesium strips to the test tube containing the acid. Measure how many seconds it takes for the magnesium to be completely consumed by the acid. Record the result in the table on the report sheet. Record any temperature changes. Dilute the unused acid which you have carefully saved with deionized water until the total volume is again 25.0 mL. Compute the new concentration and record it in the table.

Measure 12.5 mL of the diluted acid into a medium-sized test tube. **Save the remaining acid.** Place the test tube in the water bath and allow thermal equilibrium to be established. Measure how many seconds it takes for the magnesium strip to be completely consumed. Record the result. Repeat the dilution process twice more as you have done recording the new concentrations in the table. Each time allow thermal equilibrium to be established and record the time required for the metal to be consumed. Also record the temperature and any observed changes.

Predict what will happen to the time required for the magnesium strip to be consumed if you if you increase the temperature at which the reaction takes place. Write the result in the space provided. Place the bunsen burner underneath the water bath and with a low flame raise the temperature of the water bath so that it is 10 degrees hotter than room temperature. Measure in your 50 mL graduated cylinder the volume of 6.0 M HCl you have computed in the pre-lab necessary to prepare 50.0 mL of 1.0 M HCl(aq). Add enough deionized water so that the total volume is 50.0 mL. Stir the mixture thoroughly. Measure out 12.5 mL of the 1.0 M solution into each of 3 medium-sized test tubes. Place one of the test tubes in the water bath and measure the temperature of the acid in the test tube. When the temperature of the acid is about the same as the temperature of water bath record the temperature of the acid. Since the reaction generates hydrogen gas **extinguish the Bunsen burner flame before you continue.** As you did in the previous part, add the magnesium strip to the acid in the test tube and time how many seconds it takes for the magnesium to be completely consumed. Record the result. Record the temperature when reaction stops. The temperature of the reaction will be the average of the initial and the final temperatures. Repeat the experiment two more times each time raising the temperature of the acid solution by about 10 °C.

Plotting the results

The rate of the reaction is inversely proportional to the time required for the reaction. In other words the longer it takes for the magnesium strip to be consumed, the slower the reaction. Mathematically this can be expressed as rate α 1/t. On the graph paper provided plot 1/t versus [H^+(aq)]. Put a small circle in pencil around each of the data points. On the next graph plot 1/t versus [H^+(aq)]^2. Again draw a small circle in pencil around each of the data points. Now look at the two plots. Which one fits the function y =mx + b, a straight line better? What do you conclude about the reaction order with respect to H^+ concentration? Does the rate law depend on [H^+] or [H^+]^2 ?

Now we will consider the effect that temperature has on the rate of a reaction. Qualitatively, what is the relationship between the rate of a reaction and temperature? Svante Arrhenius, a Swedish chemist developed a mathematical relationship which describes how the rate constant for a chemical reaction depends on temperature. It is k= $Ae^{-E_a/RT}$. In this equation E_a is called the activation energy which is the minimum amount of energy required for a chemical reaction to take place. A is called the frequency factor which takes into account the frequency of collisions and the speed of the molecules, and T is the temperature in kelvins. We can see that this equation can be plotted as a straight line if we take the natural log of both sides which gives
ln k = -E_a/R(1/T). Since we know that rate α 1/t and rate α k then k α 1/t. Using your data fill in the table of 1/t , ln(1/t), average T in K and 1/T. You should have data for four temperatures

including the room temperature measurement you made in Part I. On the grid provided plot ln(1/t) versus 1/T. You should obtain a line with a negative slope. Using a ruler draw the best line through your data points. Pick two points on the line not too close together. Determine the slope of the line as $\Delta y/\Delta x$. This will be equal to $-E_a/R$. Here R is 8.314 J/(K·mol). Solve for E_a. Report the value of the activation energy you found.

Part II. The half-life of a chemical reaction

Procedure. Put 20.0 mL tap H_2O into a 25 mL graduated cylinder. **Record the initial volume to the nearest .1 mL .** Obtain a glass tube from the cart. Place the glass tube in the graduated cylinder so that it is centered and touches the bottom. The water levels inside and outside the glass tube will be approximately the same. Place your index finger over the top of the glass tube. With your finger still on the top of the tube remove the tube from the graduated cylinder. Transfer the water trapped inside the tube into an empty beaker by releasing your finger from the top of the tube. **Record the volume.** Repeat the transfer process 24 more times, **recording the volume of water remaining in the cylinder after each transfer.**

Calculations: Plot the volume of water vs. the transfer number . Make a small circle around each data point. Draw a smooth curve which best fits the data points.

Name	Lab Instructor
Lab Partner	Date

Experiment 16
Concepts in Chemical Kinetics

Part I. Factors affecting the rate- Concentration of H^+

Data

Trial #	M HCl(aq)	t, time for Mg to be consumed
1		
2		
3		
4		

Calculations

Trial #	$[H^+(aq)]$, M	$[H^+(aq)]^2$, M^2	1/t
1			
2			
3			
4			

Graph of 1/t versus $[H^+(aq)]$

On the graph paper plot 1/t versus $[H^+]$. Draw a small circle in pencil around each data point. Provide a title and label the axes.

On another piece of graph paper plot 1/t versus $[H^+]^2$. Draw a small circle in pencil around each data point. Provide a title and label the axes.

Does the rate law depend on $[H^+]$ or $[H^+]^2$? Explain. Then write the rate law.

Part I. Factors affecting the rate- temperature

Data

Trial #	T in °C, initial	T in °C, final	t, time for Mg to be consumed
1			
2			
3			
4			

Qualitatively describe the relationship between temperature and the rate of a chemical reaction.

Calculations

Trial #	1/t	ln(1/t)	average T in K	1/T

Graph of ln(1/t) versus 1/T

On the graph paper plot ln(1/t) versus 1/T. Be sure to provide a title and label the axes.

slope of line =
Show calculation.
Be sure to include units.

Report the value of activation energy, E_a
Show calculation
Be sure to include units

Part II: The half-life

Data

Transfer #	Volume H_2O, mL	Transfer #	Volume H_2O, mL
0		13	
1		14	
2		15	
3		16	
4		17	
5		18	
6		19	
7		20	
8		21	
9		22	
10		23	
11		24	
12		25	

On the graph paper plot transfer # on the x –axis and volume in mL on the y- axis. Provide a title and label the axes.

Questions for Part II:

1. How many transfers does it take for the volume of water in the graduated cylinder to reach 1/2 its initial volume ?

2. What will happen to the "half-life" determined in question #1 if
    a. the diameter of the tube is increased?
    b. the initial volume of water is increased.

3. Suppose the x-axis has units of time rather than transfer#. Propose a reasonable definition for half-life.

Experiment 17
Chemical Equilibrium

Introduction: One of the most important themes that runs through the second semester of General Chemistry is **equilibrium**. We learn that when a system has reached equilibrium the rate of the forward reaction equals the rate of the reverse reaction. Equally important is that equilibrium is a **dynamic** process. That is, although macroscopically the system may not appear to change as for example, in a saturated solution, on a microscopic level changes are still happening. In this experiment we will create some chemical equilibria and observe their properties.

Part I: Establishing equilibrium We will simulate the equilibrium A $\rightleftarrows$ B. There are many real systems which can be represented by this equilibrium.

Procedure: Work with a lab partner. Put 15.0 mL of tap H_2O into a 25 ml graduated cylinder. Label this cylinder A. Put 10.0 mL of tap H_2O into another 25 mL graduated cylinder and label it B. **Record the initial volumes of water on your report sheet.** Obtain two different diameter glass tubes from the cart. Look at the inside diameter of each glass tube and determine which tube has the larger diameter. Place the larger diameter tube into graduated cylinder A. Position the glass tube so that it is in the center of the graduated cylinder and one end of the tube touches the bottom of the graduated cylinder. Now place your index finger over the top of the glass tube. The level of the water inside and outside the tube will be approximately the same height. Position the other glass tube in the graduated cylinder B in the same manner as you did with A. Put your other index finger over the top of the tube. With your fingers still on the top of the tube **simultaneously** remove both tubes from the graduated cylinders. Transfer the water in each tube into the other graduated cylinder by releasing your index fingers from the top. Be careful not to spill any water. **Record the volume of H_2O in each graduated cylinder.** After you record the volumes put each glass tube back into the original graduated cylinder. Repeat the simultaneous transfer process 35 times **recording the volume of both cylinders after each transfer. After the 35th transfer stop. Do not throw out the water in the graduated cylinders. You will need it for Part II.**

Calculations: On the piece of graph paper provided, plot the volume of A in mL versus the transfer number. On the same graph, plot the volume of the other graduated cylinder versus transfer number. Draw a tiny circle around each data point. Draw a smooth curve through the data plots for A. Do the same for B. For each transfer calculate the quantity V_B/V_A where V_A and V_B are the volume in each graduate cylinder. Record your results on the report sheet.

Part II. Disturbing the equilibrium.

Procedure: Add 10 mL of liquid to the graduated cylinder having the smaller volume of water. Repeat the simultaneous transfer process 35 times as before. **Record the volumes of water for each of the two graduated cylinders.**

Calculations: Calculate V_B/V_A for each transfer. Record your results.

Part III. A real chemical equilibrium.

Procedure: Add 10 drops of NH$_4$SCN(aq), ammonium thiocyanate to a large test tube. Add 6 drops of FeCl$_3$(aq), iron(III) chloride solution followed by 20 mL of .010 M HCl(aq). Mix well with a clean glass rod. Decant about 5 mL of this solution into each of four smaller test tubes. Test tube 1 serves as a control for comparison. Add 6 drops NH$_4$SCN(aq) to test tube 2. Add 6 drops of FeCl$_3$(aq) solution to the test tube 3. Add 3 drops of 6M NaOH to test tube 4. Stir each solution. **Note all color changes compared to test tube 1.**

Name	Date
Lab Partner	Lab Instructor

Experiment 17
Chemical Equilibrium

Part I: Establishing equilibrium

Data and Calculations

Transfer#	V_A,mL	V_B,mL	V_B/V_A	Transfer#	V_A,mL	V_B,mL	V_B/V_A	Transfer#	V_A,mL	V_B,mL	V_B/V_A
0				13				26			
1				14				27			
2				15				28			
3				16				29			
4				17				30			
5				18				31			
6				19				32			
7				20				33			
8				21				34			
9				22				35			
10				23							
11				24							
12				25							

Questions for Part I

1) What happens to the volume of liquid in each graduated cylinder as the number of transfers increases ?

2) What happens to the quantity V_B/V_A as the number of transfers increases ? What does this mean about the rate of transfer of water both ways as the number of transfers increases ?

3) Suppose you had switched glass tubes so that the larger diameter tube was placed in graduated cylinder B and the smaller diameter tube were placed in graduate cylinder A. What would you predict the value of V_B/V_A to be for 35 transfers ?

4) Suppose you had performed the experiment as you have done except that you placed 10 ml water in graduated cylinder A and 15 ml of water in graduated cylinder B. What would the value of V_B/V_A be for 35 transfers ?

Part II. Disturbing the equilibrium.

Data and Calculations

Transfer#	V_A,mL	V_B,mL	V_B/V_A	Transfer#	V_A,mL	V_B,mL	V_B/V_A	Transfer#	V_A,mL	V_B,mL	V_B/V_A
0				13				26			
1				14				27			
2				15				28			
3				16				29			
4				17				30			
5				18				31			
6				19				32			
7				20				33			
8				21				34			
9				22				35			
10				23							
11				24							
12				25							

Questions for Part II

1. What happens to the volume of water in each graduated cylinder as the number of transfers increases ?

2. What happens to the quotient V_B/V_A as the number of transfers increases ?

3. Based on your results in part II of this experiment what happens to the quantity V_B/V_A when the system at equilibrium is disturbed?

4. For the present system the quantity V_B/V_A is also denoted by Q, the reaction quotient. Does Q appear to approach a limiting value for this experiment? If so, when?

Part III. A real chemical equilibrium.

Ferric ions, Fe^{3+} react with thiocyanate ions SCN^- according to
$Fe^{3+}(aq) + SCN^-(aq) \rightleftarrows Fe(SCN)_3(aq)$. The blood red color you see is caused by $Fe(SCN)_3(aq)$

Questions for Part III
1. What chemical changes did you observe when $NH_4SCN(aq)$ was added?

   How do you explain these changes?

2. What chemical changes did you observe when $NaOH(aq)$ was added?

   How do you explain these changes?

3. What chemical changes did you observe when $FeCl_3(aq)$ was added?

   How do you explain these changes?

Experiment 18
Concentration of an Unknown Acid

Introduction: This experiment will introduce you to the concept of a **titration**. Titrations have been used for a very long time in chemistry and are still very useful today in the quantitative analysis of materials. To help you understand exactly what a titration is let's begin with an example. Suppose someone has a solution of hydrochloric acid and wants to know the concentration of HCl(aq) in units of molarity. One way to figure this out is to look at the reaction of HCl(aq) and NaOH(aq). The reaction between these two substances is well understood. They react according to NaOH(aq) + HCl(aq) → NaCl(aq) + H_2O(l). The reaction tells us that for every mol of HCl, one mol of NaOH is required for the substances to completely react. Thus if we can determine the minimum number of moles of NaOH we have added to a sample of HCl to cause the reaction to be complete we can determine the number of mol of HCl in our sample. If the volume of the sample is known then we can simply divide number of moles by the volume in L and get the molarity.

This process of bringing together reactants in order to determine the concentration or amounts of reactants is called a **titration**. So, how then do we know if our reaction is complete? An **indicator** is used to signal when this happens. The indicators are often organic dyes which have distinctive colors which depend upon the pH of the solution. In order for us to know what the concentration of NaOH is we can either trust someone to determine the concentration for us or we can determine the concentration ourselves in a separate experiment.

Goal of today's experiment. You will be given a sample of sulfuric acid, H_2SO_4. H_2SO_4 is a diprotic acid, that is it can furnish 2 protons for every formula unit of H_2SO_4. The task before you is to **determine the concentration of this sample** in units of Molarity, M. To accomplish this goal you must first determine precisely and accurately the concentration of an NaOH solution. You will do that by reacting it with a known amount of a standardized acid. Then you titrate the your unknown sample with the NaOH(aq) solution to determine the concentration.

Procedure: Work with a lab partner. Obtain an unknown acid sample from the cart. **Record the unknown number of this sample on your lab report sheet. Also record the concentration of the H_2SO_4 solution in the carboy.** Next you need to prepare your equipment. Clean a buret using some soap and tap water. Be careful not to break off the tip of the buret by bumping it in the sink. Use a long buret brush to scrub the inside of the buret. Drain the buret though the tip. At this time you should check for leaks or clogs in the buret. If the buret leaks or clogs you must either fix it or replace the buret. Rinse the buret twice with tap water and then drain the buret. Rinse once more with DI water from the carboy. Then take about 5 to 10 mL of your NaOH solution and rinse out the buret with this solution. Drain the solution from the tip. Now fill the buret with the NaOH solution.

Since the buret will be used to **deliver** volumes accurately it is important that the liquid completely fills the tip of the buret. To do this point the tip of the buret upward and let liquid drain through the tip. Let enough liquid drain until the buret tip is completely filled with liquid. Point the tip back downward. You are now ready to use the buret. Repeat the procedure just described for the H_2SO_4 solution using the other buret.

Now deliver into an appropriately sized container about 20 mL of the standard H_2SO_4 solution. **Record the initial and final buret readings.** Add 2 drops of phenolphthalein indicator to the solution. **Record the initial NaOH buret reading.** The initial buret reading is usually close to zero, but it need not be exactly zero. Simply record what it is. If it is exactly zero write 0.00, not just 0 mL. Titrate the acid solution with NaOH. To do this slowly add NaOH. After each addition swirl the container. You should start to see a pink color appear and then disappear. As the pink color becomes more persistent you should slow down (add less) the addition of NaOH. When you get close to the endpoint you should only be adding a drop or even a fraction of a drop at a time. When the pink color persists for at least 30 seconds the reaction is complete. At this point you've reached the **endpoint** in the titration, the point at which the indicator changes color. **Record the final NaOH buret reading.** If you overshoot the endpoint and the solution is too dark you may add more H_2SO_4 and turn the solution clear and colorless again. Then add NaOH until you reach the endpoint. **Repeat the titration two more times. Record your data on the report sheet.**

Calculating the concentration of your NaOH solution. Based on the concentration of the H_2SO_4 solution given on the carboy and the number of mL of acid **calculate the number of mol of H_2SO_4**. Next write down the balanced molecular equation which describes the reaction between a solution of H_2SO_4 and a solution of NaOH. Use this equation to determine the **number of mol of NaOH you used to reach the endpoint. Based on the volume of your NaOH sample calculate the concentration of NaOH in units of M. Do this calculation for each of the three titrations you've done. Calculate the average value for the NaOH concentration.**

The average deviation. Let's say we calculate three values for our concentration of NaOH 0.125 M, 0.133 M and 0.139 M. The average of these values is 0.132 M. The deviation of each of these values from the average can be computed by subtracting the value from the average.
0.125-0.132 = -.007
0.133-0.132 = .001
0.139-0.132 = +.007
The average of these deviations is (.007 +.001+.007)/3=.005. Note that we've taken the absolute value of the deviations, i.e. made them all positive. We report the concentration of the solution based upon the average of the three titrations as 0.132 M ± .005.

Titrating the unknown acid. Once you've determined the concentration of the NaOH solution the next step is to determine the concentration of your unknown acid. Rinse out the buret containing the standard H_2SO_4 twice with tap H_2O and then once with DI H_2O. Drain the buret from the tip. Use 5 mL of your unknown solution to rinse out the buret once more and drain the buret. Now fill the buret with the unknown acid and fill the tip in the same way as you did before. Deliver about 20 mL of the unknown solution into an appropriately sized container **recording the initial and final buret readings.** Add 2 drops phenolphthalein indicator to the solution. **Record the initial NaOH buret reading.** Titrate the acid solution with NaOH. To do this slowly add NaOH. After each addition swirl the container. You should start to see a pink color appear and then disappear. As the pink color becomes more persistent you should slow down (add less) the addition of NaOH. When you get close to the endpoint you should only be adding a drop or even a fraction of a drop at a time. When the pink color persists for at least 30 seconds the reaction is complete. **Record the final NaOH buret reading.** If you overshoot the endpoint and the solution is too dark you may add more unknown acid and turn the solution clear and colorless again. Then add NaOH until you reach the endpoint. **Repeat the titration 2 more times. Record your data on the report sheet.**

Calculating the concentration of your unknown acid . Based on the concentration of the NaOH solution you've calculated and the number of mL of base required to reach the endpoint **calculate the concentration of your unknown acid.**
Do this calculation for each of the three titrations you've done. Calculate the average value and the average deviation. Report the results.

Name	Date
Lab Partner	Lab Instructor

Experiment 18
Concentration of an Unknown Acid

Concentration of standardized H_2SO_4 from carboy	Unknown sample #

Determining the concentration of NaOH

DATA AND CALCULATIONS	Trial 1	Trial 2	Trial 3
Initial volume H_2SO_4			
Final volume H_2SO_4			
Net volume H_2SO_4 delivered			
Initial volume NaOH			
Final volume NaOH			
Net volume NaOH delivered			
Mol H_2SO_4 Show calculation.			
Mol NaOH Show calculation.			
Concentration NaOH solution Show calculation.			
Average Deviation (AD) in concentration Show calculation.			
Ave concentration ± AD			

Determining the Concentration of the Unknown Acid

DATA AND CALCULATIONS	Trial 1	Trial 2	Trial 3
Initial volume Unknown			
Final volume Unknown			
Net volume Unknown delivered			
Initial volume NaOH			
Final volume NaOH			
Net volume NaOH delivered			
Mol NaOH consumed Show calculation.			
Mol Unknown consumed			
Concentration Unknown Show calculation			
Average Concentration Unknown Show calculation			
Average Deviation (AD) in Concentration Show calculation			
Average Conc ± AD			

Questions

1. Calculate the number of moles of acid in 20.55 mL of 0.164 M H_2SO_4.

2. 50.00 mL of an unknown monoprotic acid is titrated with 0.132 M NaOH. It takes 25.25 mL of base to reach the equivalence point. Calculate the concentration of the acid.

3. Suppose your lab partner accidentally records the value of the standard H_2SO_4 solution used in this experiment as 0.0175 M instead of 0.175 M. Assuming no other errors are made will the concentration of the unknown acid be   a) bigger by a factor of 10   b) smaller by a factor of 10   c) unaffected by the calculation **Explain your answer.**

Name		Date
Lab Partner	:	Lab Instructor

Pre-Lab Assignment for Experiment 19: Titration Curves

Introduction: This pre-lab assignment will illustrate some of the important features of titration curves and what information can be obtained from them. A titration curve is the plot of the pH of a solution such as an acid or a base as a titrant is being added to it. A titrant is the solution being added from a buret, usually an acid or a base. In this assignment, we'll consider the titration of a 25.00 mL sample of HCl of unknown concentration with 0.115 M NaOH. The pH of the solution is of course measured with a pH meter. Our goal will be to determine the pH at the equivalence point and the concentration of our sample.

What to Plot: Here's a table of sample data.

mL NaOH added	pH	mL NaOH added	pH
1	0.90	24	1.87
2	0.93	25	1.96
3	0.96	26	2.07
4	0.99	27	2.21
5	1.02	28	2.41
6	1.05	29	2.78
7	1.09	29.1	2.84
8	1.12	29.2	2.91
9	1.15	29.3	2.99
10	1.19	29.4	3.09
11	1.22	29.5	3.22
12	1.26	29.6	3.41
13	1.29	29.7	3.76
14	1.32	29.75	4.16
15	1.37	29.8	9.56
16	1.41	29.85	10.15
17	1.46	29.9	10.39
18	1.50	30	10.66
19	1.55	31	11.40
20	1.60	32	11.65
21	1.66	33	11.80
22	1.72	34	11.91
23	1.79	35	12.00

On the graph paper plot pH versus mL NaOH added. Choose the range of your axes carefully so that the plot fills the graph paper. The y-axis should be pH and the x-axis is mL NaOH.

Questions:

1. Write the chemical reaction that occurs during the titration.

2. Locate in the experiment the definition of equivalence point. Write it here.

3. Read the Plotting the Results section in the experiment and locate the equivalence point on the curve you've drawn using the method described. Mark it on your graph.

4. Using the titration curve you've just drawn determine how many mL of base were required to reach the equivalence point?

5. How many moles of base does this many mL correspond to? Show calculation.

6. How many moles of acid are contained in your 25.00 mL sample?

7. What is the concentration of your sample in M? Show calculation.

Experiment 19
Titration Curves

Introduction: In the pre-lab assignment, you constructed a titration curve for the reaction between NaOH(aq) and HCl(aq). You determined the pH at the equivalence point and the concentration of the HCl. In today's experiment, we'll replace the strong acid with a weak acid and see how the results differ. The weak acid is acetic acid, $HC_2H_3O_2$. Our goal is to determine the concentration and acid dissociation constant, K_a.

Procedure: Work with a lab partner. First, we need to calibrate the pH meter using a buffer solution. To simplify the experiment we will only carry out a one-point calibration procedure. The instructions to do this are contained on the one page handout furnished with the pH meter. Use a pH 7 buffer solution to calibrate the instrument. When the calibration is done, rinse off the electrodes with a few mL of DI water and carefully wipe them dry with a Kim Wipe or paper towel. The electrode is fragile, expensive and can be easily broken. Be careful not to break the tip. When you are done be sure to add a few mL of the buffer solution to the tip cover and replace the cover on the tip. Next, attach a buret clamp to the ring stand. Choose two burets from the cart. Clean each buret in the sink first using a long, skinny, test tube brush, some tap water and a few drops of soap solution. Be careful not to hit the tip of the buret in the sink. Drain the solution from the tip of the buret. Rinse with tap water twice and then make a final rinse with DI water. Drain the water from the tip of the buret. Secure the burets in the buret clamp

Obtain from the cart an acetic acid sample. **Record the sample number.** Put about 35 mL of the acid into a clean and dry beaker. Add a few mL of the acid to one of the burets with stopcock or pinch clamp in the closed position. Remove the buret from the buret clamp and tilt the buret back and forth so that the acid coats the walls of the buret. Then drain the acid from the tip into a beaker Dispose of the waste acid in the sink with a little running tap water. Now fill up the buret with the acid solution. Be sure the liquid level is on the graduated scale. Put about 50 mL of NaOH into a clean and dry beaker. **Record the concentration of the NaOH.** Add a few mL of the base to the other buret with the stopcock or pinch clamp in the closed position. Remove the buret from the buret clamp and tilt the buret back and forth so that the acid coats the walls of the buret. Then drain the base from the tip into a beaker. Dispose of this waste down the sink with a little running tap water. Now fill up the buret with the base solution. Be sure the liquid level is on scale. **Record the initial volumes of both the acid and the base to nearest hundredth of a mL.** Station a clean and dry 100 mL beaker beneath the buret and carefully let out 25 mL of the acid to the nearest hundredths of a mL. **Record the final volume. Calculate the volume of the acid delivered.** Move the pH meter close enough to the beaker so that the electrode supported by the bracket can easily be submerged in the liquid or removed from the liquid by carefully raising and lowering the bracket. Lower the electrode into the solution so that it is submerged in the solution. **Measure the pH and record it in the titration data table.** Raise the electrode up out of the solution. Let out about 1 mL measured to the nearest hundredth mL of NaOH into the acid. Swirl the beaker to mix the liquids. Lower the electrode back into the solution. **Record the buret reading in the appropriate column in the data table. Measure the pH and record it.** At first, the pH will change slowly because you are traveling in the plateau region of the titration curve. Keep adding 1 mL increments and swirling until you see the pH start to increase more rapidly. **When the pH increases by more than 0.3 pH unit, immediately stop and start adding only a drop of base at a time. Record the pH and the buret reading in mL after every addition of base.** This will take you through the steep portion of the titration curve. When the pH starts changing by less than 0.3 pH units after an addition you can go back to adding 1 mL at a time. **When the pH reaches 11 stop.**

Name	Date
Lab Partner	Lab Instructor

Experiment 18
Negotiating Titration Curves

Concentration of Unknown Acid

Acetic Acid Sample #	Concentration of NaOH(aq)
Volume of acid sample, mL.	

Titration Data Table:

NaOH buret reading, mL	Total mL NaOH added	pH	NaOH buret reading, mL	Total mL NaOH added	pH
Start=	0.00				

In the table above convert your buret readings into mL NaOH added. This can be done by considering that if the initial buret reading was for example 2.55 mL then to get the mL of NaOH added you will have to subtract 2.55 from all the buret readings.

Plotting the results

On the graph paper plot pH of the solution versus mL of NaOH added. Choose a range for your data so that the plot fills the graph paper. **Draw a circle in pencil around each data point.** Estimate the midpoint in the steep portion of the curve. Do this by imagining that you are looking for the point where if you were driving up the hill it would stop becoming steeper and start becoming less steep. In mathematical terms the midpoint is called the inflection point. **Mark the midpoint with an X.** Okay, so what? What's happening at the midpoint of the steep section? The X marks the equivalence point in the titration. The equivalence point is when stoichiometric amounts of acid and base have been brought together.

Concentration of the Acid

Write the chemical reaction that takes place between aqueous $HC_2H_3O_2$ solution and aqueous NaOH.

Calculate how many moles of NaOH you added to reach the equivalence point.

Calculate how many moles of $HC_2H_3O_2$ there are at the equivalence point. Based upon the number of moles of acid determine the concentration of the acid.

Determine the pH at the equivalence point.

The Acid Dissociation Constant, K_a

In the beginning of the titration the pH changes very little as base is added. As the base is added the salt $NaC_2H_3O_2$ is formed. $C_2H_3O_2^-$ is the conjugate base of $HC_2H_3O_2$. Since $HC_2H_3O_2$ is a weak acid then $C_2H_3O_2^-$ must be a relative good proton acceptor. Therefore, we have made a solution that contains a weak acid and its conjugate base. This solution is known as a buffer solution. Buffers solutions resist changes in pH when small amounts of acid or bases are added to them or when they are diluted with H_2O. We can understand how buffers work by considering the equilibrium present in the solution. The chemical reaction which describes the equilibrium present in a buffer solution can be written in generic terms as

$$HX(aq) + H_2O(l) \leftrightarrow H_3O^+(aq) + X^-(aq)$$

Here is HX is the generalized formula of a weak acid and X- is its conjugate base. The pH of a buffer solution can be calculated using the Henderson-Hasselbalch relationship

$$pH = pK_a + \log\left(\frac{[base]}{[acid]}\right)$$

where $pK_a = -\log K_a$ and the terms [base] and [acid] refer to the concentrations of the **conjugate** acid and base.

The pK_a of the acid can be determined if the log term were to disappear, i.e. become zero. What would the quotient $\frac{[base]}{[acid]}$ have to be for the log term to go to zero? Right.....It's when the quotient equals 1. At this point $pH = pK_a$

When does the quotient $\frac{[base]}{[acid]}$ become 1? Hint it's not at the equivalence point. To help you answer this remember that the base represents the conjugate base not the NaOH. The conjugate base concentration is the same as the concentration of the salt $NaC_2H_3O_2$. So when does the concentration of the salt equal the concentration of the acid. To answer this imagine that you react 1 mole of $HC_2H_3O_2$ with 1 mole of NaOH. When does the concentration of $NaC_2H_3O_2$ equal the concentration of NaOH?

Using the graph and the information from the box above determine the pK_a of the acid. Now calculate the K_a

Questions:

1. Compare the pH at the equivalence point in today's experiment with the pH at the equivalence point in the pre-lab experiment. Are they the same or different? If they are different how do you account for the difference?

2. When 50.00 mL of 0.100 M NaOH reacts with 25.00mL of 0.300 M HX, a weak monoprotic acid, the pH of the resulting solution is 3.74. Calculate the K_a of the acid.

3. In the space below sketch the titration curve for the titration of a sample of $NH_3(aq)$ as HCl(aq) is added to it. Will the equivalence point be at pH equal to 7, less than 7 or above 7. Explain.

Experiment 20
Solubility Product Constant of Magnesium Carbonate

Goal of the experiment: The purpose of this experiment is to determine the solubility product constant or K_{sp} for the chemical equilibrium

$$MgCO_3(s) \rightleftarrows Mg^{2+}(aq) + CO_3^{2-}(aq)$$

The K_{sp} can be written as the product of the equilibrium concentrations of $Mg^{2+}(aq)$ and $CO_3^{2-}(aq)$ or $K_{sp} = [Mg^{2+}(aq)][CO_3^{2-}(aq)]$ where the brackets mean concentration in units of molarity, M. The K_{sp} is a measure of a substance's solubilty. A large value for K_{sp} means the equilibrium favors the products and the substance is very soluble. A small value for K_{sp} indicates the reactants are favored and the substance is relatively insoluble.

We will establish chemical equilibrium by using a saturated solution of $MgCO_3$. A saturated solution is one in which the maximum amount of solute dissolves in a given quantity of solvent at a given temperature.

In order to calculate K_{sp} we need to know the concentrations of $Mg^{2+}(aq)$ and $CO_3^{2-}(aq)$ We will not need to measure **both** the concentrations of $Mg^{2+}(aq)$ and $CO_3^{2-}(aq)$ since these concentrations are stoichiometrically related to each other.

We will determine the $Mg^{2+}(aq)$ concentration by titrating with a solution of EDTA. EDTA is an abbreviation for ethylenediaminetetraacetic acid. The reaction is

$$Mg^{2+}(aq) + EDTA^{4-}(aq) \rightarrow MgEDTA^{2-}(aq)$$

where $EDTA^{4-}$ is one of the conjugate bases of EDTA. The species $MgEDTA^{4-}(aq)$ is an example of a complex ion. A complex ion is a species formed from a metal ion and ligands, molecules or ions which hold onto the metal ion. An indicator used to see the endpoint in this titration is EBT or Eriochrome Black T. In the presence of excess EDTA the indicator has a blue color. The concentration of the EDTA is known. By determining how much EDTA it takes to completely react with a measured volume of Mg^{2+} solution we can determine the Mg^{2+} concentration since the they are related stoichimetrically. The K_{sp} can then be calculated.

Procedure: Work with a lab partner. Put about 125 ml of the saturated magnesium carbonate solution into a clean and dry 250 mL beaker. Using your funnel and a piece of filter paper, filter this solution into another clean beaker or flask. Figure 1 shows how to filter your solution. The filtrate should be clear.

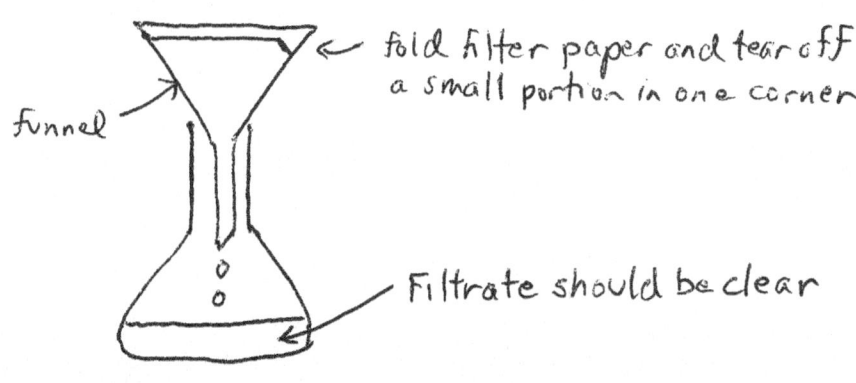

Figure 1. Filtering the solution.

While the solution is filtering, clean out two 50 ml burets. You may use tap water and a little soap. The final rinse should be with DI water.

Rinse out one of the burets with about 5 ml of your magnesium carbonate solution. Drain it from the tip. Now fill up the buret with your magnesium carbonate solution. Let out enough solution to fill the tip. The meniscus should be on scale. **Read the initial volume to 2 decimal places. Record it.**

Repeat the procedure in the preceding paragraph to fill the other buret with EDTA. **Record the Molarity of the EDTA solution. Read the initial volume on the EDTA buret. Record it.**

Use your buret to deliver about 50 ml of the magnesium carbonate solution to a clean 125 ml Erlenmeyer flask. **Record the final volume.** Add 2 drops of EBT indicator to the solution in your Erlenmeyer flask. Add about 2 ml of buffer solution to the flask. The solution should be red. Now slowly add EDTA from the other buret until the resulting solution is blue. This is the endpoint of the titration. **Record the final volume of EDTA used.**

Repeat the titration a second time recording the results in the appropriate spaces.

Calculations: Compute the net volume of EDTA solution delivered. Calculate the moles of EDTA consumed. Show your work. Look at the reaction between Mg^{2+} and $EDTA^{4-}$. How many moles of Mg^{2+} were consumed?

Compute the net volume of $MgCO_3$ solution delivered. Calculate the equilibrium concentration of Mg^{2+} in the saturated $MgCO_3$, solution. Show your work. Look at the chemical equilibrium between $MgCO_3$, and its ions. What is the concentration of CO_3^{2-} in the saturated $MgCO_3$ solution? Calculate the K_{sp}. Repeat the calculations for the second trial. Calculate an average value. Look up K_{sp} for $MgCO_3$, in your textbook. Record the value. Compare your value to the literature value. Calculate the % error.

Name	Date
Lab Partner	Lab Instructor

Experiment 20: Solubility Product Constant of Magnesium Carbonate

DATA and CALCULATIONS	Trial 1	Trial 2
1. Initial volume of $MgCO_3$		
2. Molarity of EDTA		
3. Initial volume of EDTA		
4. Final Volume of $MgCO_3$		
5. Final volume of EDTA		
6. Net Vol EDTA delivered		
7. Mol EDTA consumed. Show calculations		
8. Mol Mg^{2+} consumed		
9. Net Vol $MgCO_3$ delivered		
10. Equilibrium concentration of Mg^{2+} in $MgCO_3$ solution. Show calculations.		
11. Equilibrium concentration of CO_3^{2-} in $MgCO_3$ solution		
12. Experimental K_{sp}		
13. Average K_{sp}		
14. Accepted value of K_{sp}		
15. Percent error		

Questions:

1. Look at the bottle which contains the $MgCO_3$ solution. What evidence is there to reassure you that the solution is **saturated?**

2. Suppose the pair of students next to you secretly plot to destroy your results. They add a few ml of DI H_2O to your Erlenmeyer flask containing the filtered $MgCO_3$ solution before you titrate. What will happen to the K_{sp} value ? Will it a) increase b) decrease c) remain the same
Explain your answer.

3. Assuming the temperature dependence of the solubility of $MgCO_3$ is similar to most solids what would happen to the K_{sp} value if you performed this experiment at 10 °C instead of room temperature? Will it a) increase b) decrease c) remain the same
Explain your answer.

Experiment 21
Electrochemical Cells

Introduction: There are two basic types of electrochemical cells - those that can supply energy and those which require energy. The former are what we call voltaic cells, what we usually refer to as batteries and the latter are electrolytic cells. In terms of thermodynamic concepts a voltaic cell utilizes a spontaneous reaction whereas an electrolytic cell utilizes a non-spontaneous reaction. In this experiment we will build some voltaic cells and an electrolytic cell and study their properties.

Preliminary experiment:

Procedure: Work with a lab partner. Fill a small test tube about three-fourths full of $CuSO_4$(aq) solution and drop a strip of zinc metal into the test tube. **Record observations about this chemical change. Write the corresponding molecular and net ionic equations.** (Hint: it's a displacement reaction.) Now look at the net ionic equation you wrote. **Which species is losing electrons? Which species is gaining electrons?** In chemistry we refer to the loss of electrons as oxidation and the gaining of electrons as reduction. A reducing agent is a substance that causes another substance to be reduced. An **oxidizing agent** is a substance that causes another substance to be oxidized. **Which species is the oxidizing agent? Which species is the reducing agent?** Apart from dropping the zinc into the copper sulfate solution no additional energy was required to cause this reaction to occur. It is a **spontaneous** reaction.

Building a Daniell Cell

Procedure: Now let's put together a voltaic cell known as a Daniell Cell first built in 1812 by Frederick Daniell. Place two small clean test tubes next to each other in your test tube rack. Fill the first test tube about three-fourths full of $ZnSO_4$(aq) solution. In the second test tube put about the same amount of $CuSO_4$(aq) solution. The digital multi-meter has two wire leads attached to it. Connect a strip of zinc to one of the wire leads and place the zinc strip in the test tube containing the $ZnSO_4$(aq) solution. Connect the other wire to a piece of copper wire and place it in the other test tube. Turn on the multimeter and set it to read Volts. **Read the voltage. How many volts did you measure? (1)**

We can complete the circuit by adding a salt bridge to our arrangement. The salt bridge contains a strong electrolyte such as NaCl(aq). In a solution of NaCl there are Na^+ and Cl^- ions. These ions will migrate in the salt bridge. An old-fashioned salt bridge consists of a glass U-shaped tube containing a strong electrolyte such as NaCl. Cotton plugs are placed at either end of the U-tube to prevent the solution from rapidly mixing with the solutions in each electrode compartment. In this experiment soak a strip of filter paper in saturated NaCl(aq) solution. Immerse one end of the filter paper in the $ZnSO_4$(aq) solution and the other end in the $CuSO_4$(aq) solution. Now current can flow through our cell and we can measure a potential difference across the two electrodes. **Measure the voltage and record the voltage on your report sheet.(2)** Now switch the wires, connecting the wire lead which had been originally connected to the zinc strip to the copper wire and the wire lead originally connected to the copper wire to the zinc strip. **Read the voltage again. What do you notice? (3)**

The reaction taking place inside this voltaic cell is identical to the one in the preliminary experiment except that it only takes place when the cell delivers electricity. The electrode at which oxidation takes place is called the anode and the electrode at which reduction takes place is called the cathode. The electrons flow from the anode to the cathode. A common convention is to assign a negative sign to the anode and a positive sign to the cathode. **On your report sheet make a diagram of this cell labeling the anode and the cathode. Underneath each test tube write down the half reaction which occurs in each. Assign positive and negative signs to the electrodes.**

The main event:

Procedure: Now that you've had practice in constructing a voltaic cell we'll move on to another problem. Using iron, magnesium, and copper as materials for electrodes construct a voltaic cell which generates the **largest** voltage. To simplify this task, for each metal you should use the corresponding aqueous solution as the liquid phase of the electrode. In other words use magnesium sulfate as the aqueous phase with a strip of magnesium, use iron chloride as the aqueous phase for iron metal etc... **On your report sheet make a list of the cells you've constructed and the voltage generated for each cell.** Once you've figured out the pair of electrodes which produces the largest voltage **draw a diagram of this cell, labeling the cathode and the anode. Write down the half reaction which occurs at each electrode.**

A concentration cell:

Procedure: The voltage that a voltaic cell produces under standard conditions is denoted $E°_{cell}$. The superscript refers to standard conditions which in thermodynamics means 1 atm, any temperature (usually 25°C), and 1 M concentration. The voltage a cell produces depends upon the concentrations of reactants and products. Using the Daniell Cell you constructed change the concentrations of reactants or products to increase the voltage of the cell by a minimum of 0.05 V. Use any of the reagents available in the lab. If you need a particular reagent that is not available ask your lab instructor to see if is available from the stockroom. Explain the results you obtained using LeChatelier's principle. Hint for this part: The cell voltage will increase if the reaction will proceed to the product side as a result of an increase or decrease in concentration. A good starting place would be to examine the net ionic equation you wrote in the preliminary exercise and determine what could be done to the concentrations of either reactants or products to make that reaction proceed more to the right hand side. Then carry out those changes in concentrations.

Electrolytic cells:

Procedure: Sometimes we may wish a particular chemical reaction to occur which is non-spontaneous. We can get this reaction to occur if we supply enough energy. One way to do this is to construct an electrolytic cell.

Dry and weigh two lengths of copper wire. **Write the masses on your report sheet.** Fill a 100 mL beaker about three-fourths full of Cu^{2+} solution. Hang the copper strips on either side of the beaker so that they are suspended in the solution. **Do not let the electrodes touch each** other. Connect the positive terminal of the power supply to one copper strips and the negative terminal of the power supply to the other copper strip. Plug in the power supply. **Watch for any changes that occur.** Let the cell run for about 20 minutes. **While you're waiting read the following discussion.**

We can calculate the weight of the mass deposited at one of the electrodes with the following formula.

weight deposited in grams= $MIt/(96,500\, n)$

where M is the molar mass of the compound deposited, I is the current applied and t is the time in seconds. The number 96500 comes from Faraday's constant which is the magnitude of charge on a mole of electrons. n is the number of moles of electrons transferred per mol of substance deposited. For example in the half-reaction
$Ni^{2+} + 2e \rightarrow Ni$ the number of moles of electrons transferred is 2.

Turn off the power supply and remove the metal electrodes from your voltaic cell. Carefully dry the electrodes and then weigh them. **Record the results. Identify the metal deposited. Use the formula above to calculate how much metal deposited given the current and time you used. Compare this calculated value to your experimental results. Which electrode gains mass?**

Name	Date
Lab Partner	Lab Instructor

Experiment 21
Electrochemical Cells

Preliminary exercise:

Observations	
Molecular equation	
Net ionic equation	
Species losing electrons	Species gaining electrons

Building a Daniell Cell

1 Voltage
2 Voltage
3 Voltage

Draw a diagram of the Daniell Cell

The main event:

Cells constructed	Voltage generated

Draw a diagram of the cell which produces the largest voltage

Concentration Cell:

The voltage of the non-standard Daniell Cell is

Which reactant or product concentration(s) did you change?

Describe the method you used to change the concentration?

Explain using LeChatelier's principle why the change in concentration increased the cell voltage.

Electrolytic cell:

Mass of one electrode		Mass of other electrode	
before electrolysis	after electrolysis	before electrolysis	after electrolysis

The metal deposited is _____

Calculated mass. Assume the current is 0.50 A.

The electrode which gains mass is

Literature Cited

Laboratory Rules
The American Chemical Society, *Safety in Academic Chemistry Laboratories,* American Chemical Society, Washington, D.C. 1990

Experiment 11: Chromatography of M&Ms Candies
Mandel, M. *J. Chem. Educ.* **1992**, *69*, 988.

Experiment 12: Qualitative Analysis: Group I Cations
Bull, W.E.; Smith W.T.; Wood, J.H. *Laboratory Manual for College Chemistry*, 6th ed; Harper Collins Publishers, Inc., New York 1980; pp 235-238.

Experiment 13: Qualitative Analysis: Anions
Bull, W.E.; Smith W.T.; Wood, J.H. *Laboratory Manual for College Chemistry*, 6th ed; Harper Collins Publishers, Inc., New York 1980; pp 259-262.

Experiment 14: Solubility of Ionic Solids and Gases in Liquids
Apel, E.L.; Bannon, M.; Baron, J.; Brodemus, J.; Clevenger,E. *A Comparison of the Solubilities of Carbon Dioxide in Water at Various Temperatures,* available at http://www.woodrow.org/teachers/chemistry/institutes/1986/exp4.html
Gordon, G.; Keifer, *The Delicate Balance*, Harper and Row Publishers, New York 1980; pp.110-111.
Baer, C; Adamus, S.M. *J. Chem. Educ.* **1999**, *76*, 1540-1541.

Experiment 15: Freezing Point Depression
Bull, W.E.; Smith, W.T.; Wood, J.H. *Laboratory Manual for College Chemistry*, 6th ed; Harper Collins Publishers, Inc., New York 1980; pp 85-90.

Experiment 16: Concepts in Chemical Kinetics
Birk, J.P; Gunter, S.K. *J. Chem. Educ.* **1977**, *54*, 557-559
Carmody, W.R. *J. Chem. Educ.* **1960**, *37*, 312-313
Williams, I.W.; Hacker, R.G. *Education in Chemistry (1970) Royal Society of Chemistry available* at http://www.chemistry-react.org/filelibrary/pdf/Introtokinetics.pdf

Experiment 17: Chemical Equilibrium
Birk, J.P; Gunter, S.K. *J. Chem. Educ.* **1977**, *54*, 557-559
Bull, W.E.; Smith, W.T.; Wood, J.H. *Laboratory Manual for College Chemistry*, 6th ed; Harper Collins Publishers, Inc., New York 1980; p.105.
Carmody, W.R. *J. Chem. Educ.* **1960**, *37*, 312-313

Experiment 18: Concentration of an Unknown Acid
Bull, W.E.; Smith, W.T.; Wood, J.H. *Laboratory Manual for College Chemistry*, 6th ed; Harper Collins Publishers, Inc., New York 1980; p.105.

Experiment 19: Titration Curves
Clark, R.W.; Bonicamp, J.M.; White, G.W. *J. Chem. Educ.* **1960**, *72*, 746-750

Experiment 20: Solubility Product Constant
Skoog, D.A.; West, D.M.; Holler, F.J. *Fundamentals of Analytical Chemistry*, 5th ed; Saunders College Publishing, New York 1988; p.772